PERHAPS WOMEN

LANKES

PERHAPS WOMEN

By

Sherwood Anderson

PAUL P. APPEL, *Publisher*

MAMARONECK, N.Y.

1970

Library of Congress Catalog Number 72-105301

S.B.N. 911858-05-9

TO MAURICE LONG

INTRODUCTION

THIS little book will have to be put out as it is. It is an attempt to express, partly in story form, partly in broken verse, partly in opinions, thrust out at the readers, a growing conviction that modern man is losing his ability to retain his manhood in the face of the modern way of utilizing the machine and that what hope there is for him lies in women.

The whole thing is nothing but an impression, a sketch. I know that. I have kept it by me for a year now. I have tried to give it better form but that now seems impossible to me. I put it out hoping that it may arouse thought and discussion. I hope also it may arouse a real fear and perhaps respect for the machine, both in men and women.

At least if it arouses fear in women something intelligent may be done to save man from the dominance of the machine before his potence, his ability to save himself, is quite gone.

SHERWOOD ANDERSON

April, 1931

FRAGMENTS *from this book have been published in "Scribner's Magazine," "The Nation," "New Republic," "Household Magazine," and several European magazines. The author wishes to acknowledge his obligation to these.*

MACHINE SONG

Song written at Columbus, Georgia, in a moment of ecstasy born of a visit to a cotton mill. . . .

IT HAS been going on now for thirty, forty, fifty, sixty years. I mean the machine song. It began away back of that. I am speaking now of the great chorus, the grand song.

I am speaking of machinery in America, the song of it, the clatter of it, the whurrrr, the screech, the hummmm, the murmur, the shout of it.

It was there before the World War, before the Civil War.

The machines talk like blackbirds in a meadow at the edge of a cornfield, the machines shout, they dance on their iron legs.

The machines have a thousand, a million little steel fingers. They grasp things. Their fingers grasp steel. They grasp the most delicate cotton and silk fibres. There are great hands of steel, giant hands.

They are picking up and are handling iron pillars, great steel beams. The hands are themselves machines. They grasp huge beams of steel, swing them high up.

Steel hands are tearing up the earth. The fingers reach down through stone, through clay and muck.

They swing great handfuls of earth and stone aloft. They carry steel beams weighing tons, running with them madly across a room.

They make bridges. They make great dams. They feed upon the power in rivers. They eat white coal.

Wheels are groaning, wheels are screeching.

It is good to get these sounds into the ears. It is good to see these sights with the eyes. See the smoke rolling up, the black smoke. See the fire belching from the great retorts. The machines are cruel as men are cruel. The little flesh and blood fingers of men's hands drive, direct, control the machines.

The machines wear out as men do. Machines are scrapped, thrown on the scrap heap.

MACHINE SONG

At the edges of American cities you will see fields and gullies filled with iron and steel scraps.

There, in that gully, beneath bushes, overgrown by weeds, is an automobile that, but a year or two ago, slid smoothly over roads at forty, fifty, sixty, eighty miles an hour.

How smoothly it ran, how surely. It carried me from Chicago to Miami.

I was in Chicago and it was bleak and cold there. I wanted the sun. Cold winds blew in from the lake. My bones ached. I wanted the sun. I am no longer young. I wanted the sun.

I got into the machine. It was gaily painted. I tell you there will no man live in my day who does not accept the machine.

I myself rejected it. I scorned it. I swore at it. It is destroying my life and the lives of all of the men of my time, I said.

I was a fool. How did I know it would serve me like this?

I went to lie by the river banks. I walked in fields where there were no machines.

Is the machine more cruel than the rain?

Is the machine more cruel than distance?

Is the machine more cruel than snow?
Is the machine more cruel than the sun?

Now the machine in which I rode so gaily from Chicago to Miami, the long, graceful machine, painted a bright scarlet, now it is on a scrap heap. It is in a gully under weeds. In a few years I shall be underground. I shall be on a scrap heap.

What is worth saving of the machine, in which I rode from Chicago to Miami, passing rivers, passing towns, passing cities, passing fields and forests—what is worth saving of the machine will go into the great retorts. It will be melted into new machines. It will sing and fly and work again. What is worth while in me will go into a stalk of corn, into a tree.

I went in the machine from Chicago to Miami. Bitter winds and snow blew about me. My hands that guided the machine were cold.

It ran gaily. There was a soft murmuring sound. Something within the machine sang and something within me sang. Something within me beat with the steady rhythmic beating of the machine.

The machine gave its life to me, into my keeping. My hands guided it. With one turn of my wrists I could have destroyed the machine and myself.

There were crowds of people in the streets of some of the towns and cities through which I passed. I could have destroyed fifty people and myself in destroying the machine.

I passed through Illinois, through Kentucky, Tennessee, North Carolina and went on into Florida. I saw rains, I saw mountains, I saw rivers.

I had a thousand sensations. At night I slept in hotels. I sat in hotel lobbies and talked to men.

Today I made two hundred and fifty-eight miles.

Today I made three hundred and ten miles.

Today I made four hundred miles.

We were stupid, sitting thus, telling each other these bare facts. We told each other nothing of what we meant. We could not tell each other.

There were fat men and lean men, old men and young men. In each man a thousand sensations not told. We were trying to express something we could not express.

I am sick of my old self that protested against the machine. I am sick of that self in me, that self in me, that self in me, that would not live in my own age.

That self in me.
That self in me.
That self in me.

In my own age.
In my own age.
In my own age.
Individuality gone.
Let it go.

Who am I that I should survive?
Let it go.
Let it go.

Steady with the hand. Give thyself, man.

I sing now of the glories of a ride in a machine, from Chicago to Miami.

Miles have become minutes. If I had music in me I would orchestrate this. There would no longer be one field, one clump of trees making a wood, one town, one river, one bridge over a river.

An automobile, going at forty, at fifty, at sixty miles an hour, passing over a bridge, strikes a certain key. There is a little note struck.

Whurrrrr.

It vibrates through the nerves of the body. The ears receive the sound. The nerves of the body absorb the sound.

The nerves of the body receive flying things through the ears and the eyes. They absorb fields, rivers, bridges.

Towns, cities, clumps of trees that come down to the road.

A clump of trees comes down to the road just so in Illinois.

A clump of trees comes down to the road just so in Tennessee.

Again in Kentucky, Virginia, North Carolina.

There is a man walking in the road in Kentucky.

There is a man walking in the road in Georgia.

The car passed over a viaduct. It makes a sound.

Whurrrrr.

There are faces seen, a thousand faces. A thousand, a hundred thousand pairs of hands are grasping the steering wheels of automobiles.

I have lost myself in a hundred thousand men, in a hundred thousand women. It is good to be so lost.

Cattle, standing in fields, beside barns, in Illinois, Kentucky, Georgia, Kansas.

Bridges, rivers beneath bridges, dead trees standing solitary in fields, clumps of trees coming down to the road.

New movement.

New music, not heard, felt in the nerves. Come on here, orchestrate this.

Touch this key — a field.

That key — a sloping field.

A creek covered with ice.

A snow-covered field.

Curves in the road.

More curves.

It rains now. The rain beats against the nose of the car.

MACHINE SONG

Who will sing the song of the machine, of the automobile, of the airplane?

Who will sing the song of the factories?

We are in the new age. Welcome, men, women and children into the new age.

Will you accept it?

Will you go into the factories to work?

Will you quit having contempt for those who work in the factories?

You singers, will you go in?

You painters, will you go in?

Will you take the new life? Will you take the factories, the inside and the outside of the factories, as once you took rivers, fields, grassy slopes of fields?

Will you take the blue lights inside of factories at night as once you took sunlight and moonlight?

Will you take a new age? Will you give yourself to a new age?

Will you love factory girls as you love automobiles?

Will you give up individuality?

Will you live, or die?

Will you accept the new age?

Will you give yourself to the new age?

LIFT UP THINE EYES

IT IS a big assembling plant in a city of the Northwest.

They assemble there the Bogel car. It is a car that sells in large numbers and at a low price. The parts are made in one great central plant and shipped to the places where they are to be assembled. There is little or no manufacturing done in the assembling plant itself. The parts come in. These great companies have learned to use the railroad cars for storage.

At the central plant everything is done on schedule. As soon as the parts are made they go into railroad cars. They are on their way to the assembling plants scattered all over the United States and they arrive on schedule.

The assembling plant assembles cars for a certain territory. A careful survey has been made. This territory can afford to buy so and so many cars per day.

"But suppose the people do not want the cars?"

"What has that to do with it?"

People, American people, no longer buy cars. They do not buy newspapers, books, foods, pictures, clothes. Things are sold to people now. If a territory can take so and so many Bogel cars, find men who can make it take the cars. That is the way things are done now.

In the assembling plant everyone works "on the belt." This is a big steel conveyor, a kind of moving sidewalk, waist-high. It is a great river running down through the plant. Various tributary streams come into the main stream, the main belt. They bring tires, they bring headlights, horns, bumpers for cars. They flow into the main stream. The main stream has its source at the freight cars, where the parts are unloaded, and it flows out to the other end of the factory and into other freight cars.

The finished automobiles go into the freight cars at the delivery end of the belt. The assembly plant is a place of peculiar tension. You feel it when you go in. It never lets up. Men here work always on tension. There is no let-up to the tension. If you can't stand it, get out.

It is the belt. The belt is boss. It moves always forward. Now the chassis goes on the belt. A hoist lifts it up and places it just so. There is a man at each corner. The chassis is deposited on the belt and it begins to move. Not too rapidly. There are things to be done.

How nicely everything is calculated! Scientific men have done this. They have watched men work. They have stood looking, watch in hand. There is care taken about everything. Look up. Lift up thine eyes. Hoists are bringing engines, bodies, wheels, fenders. These come out of side streams flowing into the main stream. They move at a pace very nicely calculated. They will arrive at the main stream at just a certain place at just a certain time.

In this shop there is no question of wages to be wrangled about. The men work but eight hours a day and are well paid. They are, almost without exception, young, strong men. It is, however, possible that eight hours a day in this place may be much longer than twelve or even sixteen hours in the old carelessly run plants.

They can get better pay here than at any other

shop in town. Although I am a man wanting a good many minor comforts in life, I could live well enough on the wages made by the workers in this place. Sixty cents an hour to begin and then, after a probation period of sixty days, if I can stand the pace, seventy cents or more.

To stand the pace is the real test. Special skill is not required. It is all perfectly timed, perfectly calculated. If you are a body upholsterer, so many tacks driven per second. Not too many. If a man hurries too much too many tacks drop on the floor. If a man gets too hurried he is not efficient. Let an expert take a month, two months, to find out just how many tacks the average good man can drive per second.

There must be a certain standard maintained in the finished product. Remember that. It must pass inspection after inspection.

Do not crowd too hard.

Crowd all you can.

Keep crowding.

There are fifteen, twenty, thirty, perhaps fifty such assembling plants, all over the country, each serving its own section. Wires pass back and forth daily. The central office — from which all the parts come — at Jointville is the nerve center. Wires come in and go out of Jointville. In so and so many hours Williamsburg, with so and so many men, produced so and so many cars.

Now Burkesville is ahead. It stays ahead. What is up at Burkesville? An expert flies there.

The man at Burkesville was a major in the army. He is the manager there. He is a cold, rather severe, rather formal man. He has found out something. He is a real Bogel man, an ideal Bogel man. There is no foolishness about him. He watches the belt. He does not say foolishly to himself, "I am the boss here." He knows the belt is boss.

He says there is a lot of foolishness talked about the belt. The experts are too expert, he says. He has found out that the belt can be made to move just a little faster than the experts say. He has tried it. He knows. Go and look for yourself. There are the men out there on the belt, swarming along the belt, each in his place. They are all right, aren't they?

Can you see anything wrong?

Just a trifle more speed in every man. Shove the pace up just a little, not much. With the same number of men, in the same number of hours, six more cars a day.

That's the way a major gets to be a colonel, a colonel a general. Watch that fellow at Burkesville, the man with the military stride, the cold steady voice. He'll go far.

Everything is nicely, perfectly calculated in all the Bogel assembling plants. There are white marks on the floor everywhere. Everything is immaculately clean. No one smokes, no one chews tobacco, no one spits. There are white bands on the cement floor along which the men walk. As they work, sweepers follow them. Tacks dropped on the floor are at once swept up. You can tell by the sweepings in a plant where there is too much waste, too much carelessness. Sweep everything carefully and frequently. Weigh the sweepings. Have an expert examine the sweepings. Report to Jointville.

Jointville says: "Too many upholsterers' tacks wasted in the plant at Port Smith. Belleville produced one hundred and eleven cars in a day, with seven hundred and forty-nine men, wasting only nine hundred and six tacks."

It is a good thing to go through the plant now and then, select one man from all the others, give him a new and bigger job, just like that, offhand. If he doesn't make good, fire him.

It is a good thing to go through the plant occasionally, pick out some man, working apparently just as the others are, and fire him.

If he asks why, just say to him, "You know."

He'll know why all right. He'll imagine why.

The thing is to build up Jointville. This country needs a religion. You have got to build up the sense of a mysterious central thing, a thing working outside of your knowledge.

Let the notion grow and grow that there is something superhuman at the core of all this.

Lift up thine eyes, lift up thine eyes.

The central office reaches down into your secret thoughts. It knows, it knows.

Jointville knows.

Do not ask questions of Jointville. Keep up the pace.

Get the cars out.

Get the cars out.

Get the cars out.

The pace can be accelerated a little this year. The men have all got tuned into the old pace now.

Step it up a little, just a little.

They have got a special policeman in all the Bogel assembling plants. They have got a special doctor there. A man hurts his finger a little. It bleeds a little, a mere scratch. The doctor reaches down for him. The finger is fixed. Jointville wants no blood poisonings, no infections.

The doctor puts men who want jobs through a physical examination, as in the army. Try his nerve reactions. We want only the best men here, the youngest, the fastest.

Why not?

We pay the best wages, don't we?

The policeman in the plant has a special job. That's queer. It is like this. Now and then the big boss passes through. He selects a man off the belt.

"You're fired."

"Why?"

"You know."

Now and then a man goes off his nut. He goes fantod. He howls and shouts. He grabs up a hammer.

A stream of crazy profanity comes from his lips.

There is Jointville. That is the central thing. That controls the belt.

The belt controls me.

It moves.

It moves.

It moves.

I've tried to keep up.

I tell you I have been keeping up.

Jointville is God.

Jointville controls the belt.

The belt is God.

God has rejected me.

You're fired.

Sometimes a man, fired like that, goes nutty. He gets dangerous. A strong policeman on hand knocks him down, takes him out.

You walk within certain definite white lines.

It is calculated that a man, rubbing automobile bodies with pumice, makes thirty thousand and twenty-one strokes per day. The difference between thirty thousand and twenty-one and twenty-eight thousand and four will tell a vital story of profits or loss at Jointville.

Do you think things are settled at Jointville, or at the assembling plants of the Bogel car scattered all over America? Do you think men know how fast the belt can be made to move, what the ultimate, the final pace will be, can be?

Certainly not!

There are experts studying the nerves of men, the movements of men. They are watching, watching. Calculations are always going on. The thing is to produce goods and more goods at less cost. Keep the standard up. Increase the pace a little.

Stop waste.

Calculate everything.

A man walking to and from his work between white lines saves steps. There is a tremendous science of lost motion not perfectly calculated yet.

More goods at less cost.

Increase the pace.

Keep up standards.

It is so you advance civilization.

In the Bogel assembling plants, as at Jointville itself, there isn't any laughter. No one stops work to play. No one fools around or throws things, as they used to do in the old factories. That is why Bogel is able to put the old-fashioned factories, one by one, out of business.

It is all a matter of calculation. You feel it when you go in. You feel rigid lines. You feel movement. You feel a strange tension in the air. There is a quiet terrible intensity.

The belt moves. It keeps moving. The day I was

there a number of young boys had come in. They had been sent by a Bogel car dealer, away back somewhere in the country. They had driven in during the night and were to drive Bogel cars back over country roads to some dealer. A good many Bogel cars go out to dealers from the assembling plants, driven out by boys like that.

Such boys, driving all night, fooling along the road, getting no sleep.

They have a place for them to wait for the cars in the Bogel assembling plants. You have been at dog shows and have seen how prize dogs are exhibited, each in his nice clean cage. They have nice clean cages like that for country boys who drive in to Bogel assembling plants to get cars.

The boys come in. There is a place to lie down in there. It is clean. After the boy goes into his cage a gate is closed. He is fastened in.

If a country boy, sleepy like that, waiting for his car, wandered about in a plant he might get hurt.

There might be damage suits, all sorts of things.

Better to calculate everything. Be careful. Be exact.

Jointville thought of that. Jointville thinks of

everything. It is the center of power, the new mystery.

Every year in America, Jointville comes nearer and nearer being the new center. Men nowadays do not look to Washington. They look to Jointville.

Lift up thine eyes, lift up thine eyes.

LOOM DANCE

THEY HAD brought a "minute-man" into one of the Southern cotton-mill towns. A doctor told me this story. The minute-men come from the North. They are efficiency experts. The North, as everyone knows, is the old home of efficiency. The minute-man comes into a mill with a watch in his hand. He stands about. He is one of the fathers of the "stretch-out" system. The idea is like this:

There is a woman here who works at the looms. She is a weaver. She is taking care, let us say, of thirty looms. The question is — is she doing all she can?

It is put up to her. "If you can take care of more looms you can make more money." The workers are all paid by the piece-work system.

"I will stand here with this watch in my hand,

You go ahead and work. Be natural. Work as you always did.

"I will watch every movement you make. I will coordinate your movements.

"Now, you see, you have stopped to gossip with another woman, another weaver.

"That time you talked for four minutes.

"Time is money, my dear.

"And you have gone to the toilet. You stayed in there seven minutes. Was that necessary? Could you not have done everything necessary in three minutes?

"Three minutes here, four minutes there. Minutes, you see, make hours and hours make cloth."

I said it was put up to her, the weaver. Well, you know how such things are put up to employees in any factory. "I am going to try this," says the boss. "Do you approve?"

"Sure."

What else is to be said?

There are plenty of people out of work, God knows.

You don't want to lose your job, do you?

(The boss speaking.)

"Well, I asked them about it. They all approved.

"Why, I had several of them into my office. 'Is everything all right?' I asked. 'Are you perfectly satisfied about everything?'

" 'Sure,' they all said."

It should be understood, if you do not understand, that the weaver in the modern cotton mill does not run his loom. He does not pull levers. The loom runs on and on. It is so arranged that if one of the threads among many thousand threads breaks, the loom automatically stops.

It is the weaver's job to spring forward. The broken thread must be found. Down inside the loom there are little steel fingers that grasp the threads. The ends of the broken thread must be found and passed through the finger that is to hold just that thread. The weaver's knot must be tied. It is a swiftly made, hard little knot. It will not show in the finished cloth. The loom may run for a long time and no thread break, and then, in a minute, threads may break in several looms.

The looms in the weaving-rooms are arranged

in long rows. The weaver passes up and down. Nowadays, in modern mills, she does not have to change the bobbins. The bobbins are automatically fed into the loom. When a bobbin has become empty it falls out and a new one takes its place. A full cylinder of bobbins is up there, atop the loom. The full bobbins fall into their places as loaded cartridges fall into place when a revolver is fired.

So there is the weaver. All she, or he, has to do is to walk up and down. Let us say that twenty or thirty looms are to be watched. The looms are of about the breadth of an ordinary writing desk or the chest of drawers standing in your bedroom.

You walk past twenty or thirty of them, keeping your eyes open. They are all in rapid motion, dancing. You must be on the alert. You are like a school teacher watching a group of children.

But these looms, these children of the weaver, do not stand still. They dance in their places. There is a play of light from the factory windows and from the white cloth against the dark frames of the looms.

Belts are flying. Wheels are turning.

The threads — often hundreds to the inch —

lie closely in the loom, a little steel finger holding each thread. The bobbin flies across, putting in the cross threads. It flies so rapidly the eye cannot see it.

That is a dance, too.

The loom itself seems to jump off the floor. There is a quick jerky movement, a clatter. The loom is setting each cross-thread firmly in place, making firm, smooth cloth.

The dance of the looms is a crazy dance. It is jerky, abrupt, mechanical. It would be interesting to see some dancer do a loom dance on the stage. A new kind of music would have to be found for it.

There are fifteen looms dancing, twenty, thirty, forty. Lights are dancing over the looms. There is always, day in, day out, this strange jerky movement, infinitely complex. The noise in the room is terrific.

The job of the minute-man is to watch the operator. This woman makes too many false movements. "Do it like this."

The thing is to study the movements, not only of the weavers but of the machines. The thing is to more perfectly coordinate the two.

It is called by the weavers the "stretch-out."

It is possible by careful study, by watching an

operator (a weaver) hour after hour, standing with watch in hand, following the weaver up and down, to increase the efficiency by as much as 100 per cent. It has been done.

Instead of thirty-six looms, let us say seventy-two. Something gained, eh? Every other operator replaced.

Let us say a woman weaver makes twelve dollars a week. Let her make sixteen. That will be better for her.

You still have eight dollars gained.

What about the operator replaced? What of her?

But you cannot think too much of that if you are to follow modern industry. To every factory new machines are coming. They all throw workmen out of work. That is the whole point. The best brains in America are engaged in that. They are making more and more complex, strange and wonderful machines that throw people out of work.

They don't do it for that reason. The mill owner doesn't buy for that reason. To think of mill owners as brutes is just nonsense. They have about as much chance to stop what is going on as you have.

What is going on is the most exciting thing in modern life. Modern industry is a river in flood, it is a flow of refined power.

It is a dance.

The minute-man the doctor told me about made a mistake. He was holding his watch on the wrong woman.

She had been compelled to go to the toilet and he followed her to the door and stood there, watch in hand.

It happened that the woman had a husband, also a weaver, working in the same room.

He stood watching the man who was holding the watch on his wife in there. His looms were dancing — the loom dance.

And then suddenly he began to dance. He hopped up and down in an absurd, jerky way. Cries, queer and seemingly meaningless cries, came from his throat.

He danced for a moment like that and then he sprang forward. He knocked the minute-man down. Other weavers, men and women, came running. Now they were all dancing up and down. Cries were coming from many throats.

LOOM DANCE

The weaver who was the husband of the woman back of the door had knocked the minute-man down, and now was dancing upon his body. He kept making queer sounds. He may have been trying to make the music for the new loom dance.

The minute-man from the North was not a large man. He was slender and had blue eyes and light, curly hair and wore glasses.

The glasses had fallen on the floor.

His watch had fallen on the floor.

All the looms in the room kept running.

Lights danced in the room.

The looms kept dancing.

A weaver was dancing on a minute-man's watch.

A weaver was dancing on a minute-man's glasses.

Other weavers kept coming.

They came running. Men and women came from the spinning-room.

There were more cries.

There was music in the mill.

And really you must get into your picture the woman — in there.

We can't leave her out.

She would be trying, nervously, to arrange her clothes. She would have heard her husband's cries.

She too would be dancing, grotesquely, in a confined place.

In all the mills, the women and girls hate more than anything else being watched when they go to the toilet.

They speak of that among themselves. They hate it more than they hate long hours and low wages.

There is a kind of deep human humiliation in that.

There is this secret part of me, this secret function, the waste of my body being eliminated. We do not speak of that. It is done secretly.

We must all do it and all know we must all do it. Rightly seen it is but a part of our relations with nature.

But we civilized people are no longer a part of nature. We live in houses. We go into factories.

These may be a part of nature, too. We are trying to adjust ourselves. Give us time.

You — do not stand outside of this door to this

little room, holding a watch in your hand, when I go in here.

There are some things in this world, even in our modern mass-production world, not permitted.

There are things that will make a weaver dance the crazy dance of the looms.

There was a minute-man who wanted to co-ordinate the movements of weavers to the movements of machines.

He did it.

The legs of weavers became hard and stiff like legs of looms. There was an intense up and down movement. Cries arose from many throats. They blended strangely with the clatter of looms.

As for the minute-man, some other men, fore-men, superintendents and the like, got him out of there. They dragged him out at a side door and into a mill yard. The yard became filled with dancing, shouting men, women and girls. They got him into another machine, an automobile, and hurried him away. They patched him up. The doctor who patched him up told me the story.

He had some ribs broken and was badly

bruised, but he lived all right. He did not go back into the mill.

The "stretch-out" system was dropped in that mill in the South. The loom dance of the weavers stopped it that time.

IT IS A WOMAN'S AGE

IT'S A woman's age. Almost any man will admit that. He admits it rather sadly.

To my mind all sorts of things are involved in the matter. For the present at least men are somewhat lost.

In America it is an age without religion. Who would dare venture the assertion that the Americans are, at present, a religious people? They have taken old beliefs from us, have taken the life out of them, and have given us no new ones.

Something a bit queer has happened to man. The age has moved too fast for him. Science has succeeded in killing most of the old mystery. Who dare question the assertion? The machine has taken from us the work of our hands. Work kept men healthy and strong. It was good to feel things being done by our hands. The ability to do things to materials with our hands and our heads gave us a certain power over women that is being lost.

There are too many goods. The idea that it is man's noblest purpose to supply the world with goods has been carried too far. The modern man is drowned in a flood of things he did not make.

He has no definite connection with the things with which he is surrounded, no relations with the clothes he wears, the house he lives in. He lives in a house but he did not build it. He sits in a chair but he did not make it. He drives a car but he did not build it. He sleeps in a bed but he does not know where it came from.

He goes too easily from place to place. Places begin to mean less and less to him. He cannot remember the places. How many men are like myself, going restlessly from place to place, seeking something they cannot find? Places also have lost their significance. There is no mystery in distance. The more obvious mysteries of life have been destroyed too fast.

The scientists have taken from us old mysteries, and, as yet, no poets have arisen who can give us new ones.

It is the nature of man to need God, the mystery. Without the mystery we are lost men.

You have to admit it is not a laughing age. Listen in on the radio. All the laughter is fake laughter. It doesn't go down through the body to the toes. Notice how little real laughter you hear on the streets.

Listen to the voices coming over the radio. There is no reality to the voices. Who can speak naturally over the radio? The little machine, standing coldly there, is too much for the man. It confuses him, makes him ashamed. Dimly he realizes its possibilities. They are overwhelming.

Something creeps into the voices coming over the radio. They sound false.

I am myself convinced that all this has had a profound effect upon the relationship of men and women. I think that is pretty obvious.

It did not start there. It may have started in the factories. When more men worked in the fields and when most of the goods we need to cover our nakedness against the cold, the houses we live in, were made by men's hands, men were different.

They believed more fully in the mystery of existence. That fact gave man a certain dignity. He was, at the same time, more sure of himself and more humble. Most of the modern assertiveness of man is due to fear.

What is there for me to be proud about in the telephone, in the flying-machine, in the radio? What had I to do with them?

It is a bit absurd for me to try to make anything. I had better just sit dumbly here. The machine can make whatever I need in the common affairs of my life faster and better than I can.

My own notion is that men need this direct connection with nature in work. They need to touch materials with their hands. They need to form materials, need to make things with their own hands out of wood, clay, iron, etc. They need to own tocls and handle tools.

Not doing it, not being permitted to do it, does something to men. They all know it. They hate to admit it, but it is true. Not being able to do it makes them less men. They become no good for women. They spoil things for women too.

It is because of the factories.

But wait! The factories are themselves all right. The big complicated beautiful machines in the factories are in no way to blame. They are gorgeous things. I have been spending some months in factories now, just looking at the machines. Few enough people, not compelled to work in the factories, ever go into them, but they should go.

The machines are beautiful with a cold kind of classic beauty, but they are beautiful. In motion they become gorgeous things. I have stood sometimes for two or three hours in some big factory looking at the machines in motion. As I stand looking at them my body begins to tremble. The machines make me feel too small. They are too complex and beautiful for me. My manhood cannot stand up against them yet. They do things too well. They do too much.

I have to keep telling myself over and over, "wait," I have to keep telling myself, "remember men made these machines."

It may be that the men who make the machines are all right. They may be healthy. I can think of the man who makes machines, with a kind of gladness in me. He has an idea in his head. So he gets so much iron, steel, wood, what-not. The idea he has in his head works down through his hands. It is like my writing these words on this paper here.

The ideas form in my head and then automatically my hands try to express the ideas in words.

This is what a machine-builder may do with his materials. He has certain problems before him. If he is, for example, building a machine to fly through the air, he has to economize. He can't waste materials, make the machine too heavy. He can't do that in building any machine. He has to avoid waste of materials.

That leads to efficiency and beauty. We have got to that in the machines. Most of the basic principles connected with the manufacture of goods in machines have been found out now. We are refining now, we are making the machines more and more complex, we are fabricating more.

After all, the machine is only a tool, but for the present, at least, it is too big, too efficient for us.

It plays the very devil with the man who works the machine. The man who works the machine feels too small. He is working all day and every day in the presence of something apparently stronger than himself, more efficient. It makes him feel inferior. His spirit gets tired. The spirit of the machine doesn't tire—it hasn't any.

The real problem of what the machines are doing to men hasn't begun to be touched yet.

But what about the women? Why are they more triumphant than the men in such an age? Perhaps they aren't. It may be they only seem to be less touched.

There are more goods made. Women get most of the goods. Men earn more money. The women spend it. They do right. Women are more practical than men. If they can't get what they want, they will take what they can get.

If they cannot get men, they will take goods.

But what do they want?

Love, eh?

Where are they to get the lovers, the mates?

There are more and more women going into the factories to work. I have been going into factories and standing in factories. For months now I have been watching the machines. I have been watching men and women who run the machines. I get nothing out of anything in life except as it affects the lives of men and women. There is a sense in which I have hated these machines I have seen in factories, but only because I am a man and hate to see my fellow men overawed, crushed, humiliated, made to feel small.

In the factories the men employees seem to feel smaller than the women. The women are affected less. It must be because every woman has a life within herself that nothing outside her can really touch except maybe a mate. It can be a very still, beautiful, waiting life. The women who work in the factories do not seem to want the men who work in the factories as mates. They seem to have a kind of contempt for them. Girl clerks who work in department stores do not want the men who work in department stores.

No woman really wants a man who feels defeated, crushed by life.

You look at her like this — the woman — there she is! Let's say her father is a doctor or a lawyer or even a judge. Or he may be a business man, a sales agent for some large machinery manufacturing company, or a farmer.

There she is. Let us say she is twenty-seven or eight. She has never married. Some women marry, others do not.

Let us admit that marriage, having children,

keeping a house for a man, is a career. It is something to be done.

But we, in America, are coming more and more to the sophisticated point of view as regards marriage. It is a job. If you are a woman and have the chance, that is to say a chance that looks good enough to you, interesting enough, you take it on. So you are in love, eh?

It doesn't last.

Mating, if it were real mating, would last.

We are all coming to realize slowly, as we get more and more education in life, that love, meaning desire, comes and goes.

Mating is another thing.

You are a woman and married and suddenly you desire a man, not your husband, the father of your children, but another one.

Why? We ought to ask ourselves that question. What has life done to him to make him less the desirable lover?

Or it happens to a man. How many men who hold positions in offices are in love, or half in love, with their stenographers?

Is this new woman more lovely than the woman I have married?

She isn't with me all the time. She comes here to this office of mine from a strange house. We sit closely together for hours.

She has no hold on me. I do not have to explain myself to her. She isn't onto me? But what do I have to conceal?

Let us say the woman I have married has been foolish. She has kept nothing of herself hidden from me.

Like me, she has ugly times. Look at the comic strips in our newspapers. At least half the jokes are about men being humiliated by their wives. We all look at these strips and enjoy them. There is another poor devil being made a fool of. He is being humiliated by his wife.

"I am not the only one."

There is that absurd figure in the comic strip, "Bringing Up Father." The man in that comic strip evidently began life as a laborer. He was shrewd enough to make money.

Obviously men do like him. He is always trying to get away from his wife, to go down to Dinty Moore's. He is a welcome figure down there.

His wife, always triumphant, is as ugly a figure as the artist can draw. How the artist must hate her. She spends father's money recklessly for anything that comes into her head. The daughter does the same.

In their home, a very rich, elaborate house, father is always being brutally snubbed; he is walloped over the head with a broom, with a chair, with a milk bottle.

He is thrown out of up-stairs windows, locked in his room.

The poor man, who has obviously made the money being spent in every conceivable way by this wife and daughter, wants little. He wants to go to Dinty Moore's and play cards. He isn't young any more but he likes pretty girls, likes to look at them on the street. He likes to eat corned beef and cabbage.

It is for this reason he is punched and beaten in this terrific manner by these brutal women, day after day, month after month, year after year.

Other figures in other comic strips getting it too. I have asked several newspaper editors and publishers about all of this. "Don't your readers ever get weary of it?" I have asked.

"I mean of the men always being pounded in this absurd way by the women, by the wives? Why is it always so one-sided? Don't your readers ever get sick of it?"

"We don't dare to stop it," the newspaper editors say; "we would lose half our subscribers."

"But why not, now and then, let a husband beat up his wife — not in fact, you understand — in the comic strips?"

"It wouldn't do; they wouldn't stand for it."

"You mean the women wouldn't stand for it?"

"No, I mean the men wouldn't."

"They want to see some other poor devil getting his?"

"Yes, sure."

A pretty humiliating statement from a male point of view.

It comes and goes, this feeling between men and women. Marriage is a long, long test. Some men and women will stand for it, others won't.

If they stick it out, they gain something. Who has not seen sweet old couples? I have seen them. I have seen old Negro couples, old white couples, old Italian couples, old German couples, who have stuck together a long, long time.

Now and then you see a picture of some old

couple in a newspaper: "Mr. and Mrs. A. K. Smith Celebrate Their Golden Anniversary."

My God!

Dear old warriors.

They sit together in a house after fifty years of it.

"You, John, sit still over there. Let me take another look at you."

"You, Mary, Lord, I have lived with you for fifty years."

Most of my own class in America, the artists, the poets, the painters, singers, sculptors, music-makers, actors, storywriters, have been married more than once.

They have split off from this one and have got another one.

Sometimes a third, a fourth, a fifth.

A restless seeking after perfection that is beauty in the world of fact, the world of flesh.

"Now I have it.

"I have it not.

"I have it.

"I have it not."

You meet such a man friend. He is going

through it. He has left one woman and gone to another.

Women do it too. They leave one man and go to another.

What percentage of the men and women who go out to Reno to be divorced do you think intend to marry again? Once I asked a judge who had granted many divorces about the matter.

"What per cent of them intend to marry again, and almost immediately?"

"One hundred and twenty-five per cent," he said.

All of the above as thoughts in the head of a man — myself. Surely I am a man of a certain class. I am a man of the artist's class.

As regards human relationships, I am of course muddle-headed.

How could I be anything else, being both an artist and an American?

I swear I am both.

In a land where women rule.

In a land where, in industry, the machine rules.

It is inevitable, I think, that I must go deeper and yet deeper to get at what I am trying to say here.

I am trying to proclaim a new American world, a woman's world.

The newspapers are all run for women, the magazines, the stores.

The cities are all built for women. Whom do you suppose the automobiles are built for?

Practically all the American men I know have surrendered to women.

Are women then, in America, superior to men?

Not really. At present they have the best of it.

Because present-day life humiliates men more directly. More has been taken from them.

Here is a curious idea. For days I have been going about saying to myself something like this: "Suppose," I have said to myself, "the whole modern age, the industrial age, the machine age, were women-made."

I do not mean that woman is more inventive than man.

Man invents in the world of the imagination. When he builds a new machine, he is only bringing over into the world of fact the thing born in the imaginative world.

That is man's world, the imaginative world, and in a curious way he has surrendered it also.

It is a factual age, and in a factual age women will always rule. In the world of fact every woman has the advantage of me because she has something I cannot have — the machine cannot touch her mystery — but let her come over into my own male world, the world of fancy, and surely I will lose her there. I will go sure-footed through dim, far reaches of the fancy where she must always stumble blindly.

In America there is an obvious danger. It is that presently man, having stayed for so long a time outside of his own male world, will lose forever the sense of it.

I, myself, feel this danger constantly in the presence of modern industry. I admit that man invented all of these machines. He is still inventing. The machines are running in the factories. Great rivers of goods flow from them.

It seems to me that the goods are mostly for women. Women are the great consumers. They have a passion for possession. The passion for possession is feminine.

Your acquisitive man is a man become feminine. How can this help being true? To be rich is to become conservative, to value possessions.

Your male should be the adventurer. He should be careless of possessions, should throw them aside.

His true place is in the imaginative world, the world of fancy.

I think we modern men have lost maleness because we have not really dared face the machine, or the significance of a capitalistic civilization. I will admit that we cannot get away from the machine. We will have to get away from it all we can.

How is it to be done? I have asked myself that question over and over. We will have to keep the machine and control it. Men who must work the machines, who do not build them, should be asked to stay in the presence of the machines for the present but a few hours at a time. They should be given the chance to get strength to free themselves. They should not be asked to stay until they are overawed, humiliated too much. We will have to have frequent shifts of men in the factories.

It may sound childish, but men will have to go back to nature more. They will have to go to the fields and the rivers. There will have to be a new religion, more pagan, something more closely connected with fields and rivers.

There will have to be built up a new and

stronger sympathy as between man and man. We may find the new mystery there.

We will have to quit being afraid of sex. It isn't so terrible really. Life would lose three-fourths its charm without sex.

We will have to rediscover the wonder of our own maleness or the women will have no lovers, no mates. There will be no lovers. There will only be husbands. If something of the sort isn't done, if money and the machine continue to rule men's lives, then we shall have to surrender maleness. We will have to live like the bees in a feminine world, with a few drones flying about in the air, with perhaps just enough maleness loose in the world to keep up the show.

PERHAPS WOMEN

I WAS thinking of women one evening in the winter of 1929-1930 as I walked in the streets of a South Carolina mill village.

I was thinking of women and of modern industry in America, the coming of the factories. The factories had brought about a profound change in all life. Everyone admitted it. Most people called the change good, they called it progress. Was it that, was it really progress?

"We are in a time of change. We are in a time of change. Who can tell?" I kept saying to myself. There was evidence of a changed attitude toward life on all sides. Men were in new relation to each other, women were growing into a new relation to men.

"A time of change. A time of change," I kept saying the words over and over. The words were a song in me. It had been a grey evening after rain but the night was clear. I remember that I went stumbling along.

"A time of change. A time of change. Old values are being destroyed. Are men men enough to make for themselves new values? Will women have to do it for them?"

I was thinking these thoughts and others, wandering through the streets of a quiet miserable Southern mill village late at night. I had been given permission to visit the mill at night, see the night shift at work.

The permission had not been given too readily by the mill owners. There had been telephone conversations, people seen, visits made. "What did I want?" Most of the American radicals I have seen and to whom I have talked — the men at war with our modern life — are great romancers. They like to think of the men who own factories, the successful industrialists of our day, as devils, as Goliaths, as supermen. Perhaps to think of these men so gives some of the younger radicals a satisfying feeling of virtue in themselves.

"You see how kind I am. I love the poor, the down-trodden. What nonsense," I had said to myself; "we are all alike . . . at bottom we want to save ourselves."

If I can make another, unlike myself, with different talents, seem mean, that makes me, being different, seem to myself rather noble.

Why, it isn't as mean as that. All I am trying to say is that the rich and successful man has his confusion too. In America most of the rich and successful with whom I have talked at all intimately are also confused. They are self-conscious and puzzled.

Let us admit that the factory owners when I went to them were suspicious of me. A woman, the wife of such a man, was driving me in her car. "The factory owners, men like your husband, are suspicious of writers," I said. "They are. They have a right to be," she replied.

Is it true that the man who devotes himself to acquisitions, who becomes, perhaps by a lifetime of effort, an owner of lands, stocks and bonds, who controls factories, less interesting, as a man, as an individual, than your ragtag fellow who owns nothing?

Well, he is. Let us admit this. All literature proves it to be true.

We writers have always sought out the rascals or we have sought out the poor, the unsuccessful.

"It is only when you are torn from your mooring, when you drift like a rudderless ship, that I am able to come near to you."

I had myself written the above lines years before.

It was Mr. Henry Adams who said, "A friend in power is a friend lost." The saying is quite true. How many friends I have lost in that way! They succeeded in life, became successful men.

The rich and the successful are inevitably tied to the possessions they have acquired. They are afraid of change. If men's attitude toward possession is ever to undergo a change these men will be the enemies of change. Suppose men were trying, stumbling, to meet a new situation brought about by the coming of industrialism, mass production, etc.

More and more men thinking about it, trying to think their way through it. There is little enough doubt where the opposition will lie.

In America we have got fixed in our laws, in all our ways of thinking, the notion of the sacredness of property.

What a silly notion really. Only life should be sacred.

These thoughts in my head.

I had gone in to see a man. He was in control of a factory. A red-haired alert capable man he seemed. "I want to go into your factory at night," I said. "You are working women and girls all night long.

"They are in there, in the long night hours. The walls of the factory impinge on them.

"I want to go in there. I want to stand about. I want, if I can, to feel what they feel. I want to write about it.

"Now you look here, Mr. Factory Owner, I am interested in something else besides the stock questions now stirring the industrial world. I am not primarily interested in shorter hours or more wages. I am not trying to unionize your factory.

"I will not even speak to one of your employees. Let your superintendent take me in. I want to go in the middle of the night, at the zero hour, let us say, at two o'clock in the morning, or better yet at three o'clock.

"I am also interested in modern machinery. Machinery has become a part of all modern life. I myself run an automobile. I love driving it.

"I would like, if I live long enough, to write things that run as smoothly as modern machinery has been made to run.

"I want to give myself to the machine. I am, it is true, interested in the human side of the modern factory but that is but a part of my interest.

"The machine must be doing something to people too. It must be doing something to you and me as well as to the employees of your factory.

"That should be interesting to you as well as to me. You are the owner of the factory, it is true, but you are also a man.

"I would like to find out something."

A mill owner or manager, shaking his head.

"But you writers — why are you always against us?"

"It is all nonsense. You are a man, are you not?" I said again.

"I will admit that, as an individual, you are perhaps less interesting than a poor man, one of your employees.

"Life will have beaten down upon such a man more. He will be less sure of himself.

"That would make him more open to new impressions, new impulses.

"New forces will have to begin moving, operating on men here.

"You successful men are so walled-in by your possessions. You cannot receive new impressions."

"Oh, go to the devil."

"May I go into your factory at night, at three o'clock in the night, when the night shift is on, when everyone is a bit tired?

"It is only when you are torn from your mooring, when you drift like a rudderless ship, that I am able to come near to you."

"It can't be you are afraid of me."

"To the devil with you. Well, I'll see. I would have you understand I am not afraid. I'll see.

"Yes, damn you, you may go in."

"We are in a time of change," I kept saying to myself. "There are evidences of it on all sides. Life is shifting, changing. Old values are being destroyed.

"They have been destroyed wholesale in Russia. Will the same thing happen here?"

You, the reader, must think of me as having these thoughts, walking the streets of that Southern cotton mill village at night.

Will there be some day a revolution here?

Revolutions, when they come, are like wars. They accomplish, if they do accomplish, at terrible cost. It is a whoopla time, everything promised, sometimes nothing gained.

New cruelties often to take the place of old. New men in power. Power, the desire for power, is a disease. It destroys every attempt men make to advance.

It is a dead time. Men are standing still. They dare not face the realities of life.

Look what we Americans had seen in our time. We had seen fortunes pile up in a way never dreamed of by our fathers. Mass production had come, as an inevitable result of industrialism. We had got, in my time, a new attitude toward labor. Never perhaps in the history of the world — (but let us not be too far-reaching, let us say, "in the history of America")—never had labor sunk so low—

A time of little men when big men are so needed. What has made them little?

Is it going away from earth, sky, women, direct touch with earth and the materials that come out of earth — wood, iron, stone — through tools?

The modern world taking tools and all control out of men's hands, the instruments by which men felt their way into life through wood, iron, stone, into the lives of women too?

The woman's life being always different, the approach a different one?

Men had learned to put their faith in two things, in goods, the production of goods rapidly and at low cost, and in a thing called "publicity."

This later had become the central, the true faith. It had been carried to ridiculous lengths. Even as I write these words we are in a time of stress. Again too many goods have been produced. This will keep happening. Why not?

Have I not myself seen how every year machinery becomes more and more efficient? Does not efficiency in machinery mean less men employed? If men are not employed how are they to receive wages? You see it is no longer a question of how much wages or of how long hours but of whether or not men are to be employed at all.

And if men are not employed, if they receive no wages how are they to buy goods?

More and more goods, with less and less people employed. That is prosperity.

Why, how can you manage that?

By publicity.

By publicity.

Advertise.

Advertise.

What a strange childish faith! Even as I write these words there are millions of men out of employment and only last night a young mill engineer walked with me. We talked. The young mill engineer, one of the type of American young men on whom our faith is placed now, a kind of young Lindbergh, all machine, rather nice too, tied with all his faith to the machine, instinctively responsive to the machine—a so-called clean young man, married, leading what the newspapers call an exemplary life—such a young man walked and talked with me only last night. He grew a little excited, even poetic. There was a moon shining. Behold, even to the young Lindbergh, the moment of poetry came. Was the young Lindbergh not all poetry once — for a day and a night, out there alone over the Atlantic — the American eagle come to life — flown off the dollar for the time?

The young man who walked with me, just another young Lindbergh, was an engineer in a great American cotton mill. He dreams of what — of some day seeing a mill built that would employ no people at all. "Well," he said, "perhaps there will be a few men. They will be highly paid specialists. They will stroll through the mill, listening and looking. The cotton will come in at one end of the mill and cloth flow out at the other. No human hand will touch it."

Do you think it an impossible dream? If you do you know little of the tremendous progress already made in machine building, you know little of the American genius.

And even as I write these words there are three millions of men out of employment and no lack of goods. Men are walking the streets of industrial towns and cities out of work. Of course, they cannot buy goods.

How are such conditions to be met? Call a meeting of great industrialists!

What shall we do? Really it is very simple. Say times are good. Say that anyone who declares times are not good is un-American, a dangerous character. Call him a Bolshevik. How silly! There were men here once who were not afraid to face things, who were not frightened by plain words said. How about Thomas Jefferson, Andrew Johnson, Andrew Jackson, Abraham Lincoln?

Why only yesterday a letter came to me from a factory worker. "We are weakened and degenerated men already," he said. "I am ashamed to write these words to you but they are true. It is rare to find anyone in a position of power in America now with any interest in us. Our own leaders are not interested and it is a sign of our degeneration as men that we stand for such leadership. We are, in truth, abandoned ones. A European peasant, bronzed, sturdy and independent, as compared with us, seems like a vastly superior type of man—we creatures who have won only the contempt of our employers.

"I do not blame our employers," he wrote.

Words. Declare we are are a prosperous people. Keep declaring it. Have all of the newspaper editors declare it loudly. Have all of the leaders of industry give out interviews. A strange faith in words. We writers should not object to that faith. Most of us get on very well. All we have to do is to write whoopla.

All of the above thoughts passing through my head at night as I walked in a Southern mill village. It was a rolling red country. It was winter. That very afternoon I had been driving with a woman friend through the country about the town I was in.

But perhaps I had better call it a city. A good many thousand people lived there. There were other people besides mill employees, factory employees.

There were doctors, lawyers, judges, keepers of stores, drivers of trucks, clerks, young woman stenographers, there was a whole middle-class world.

The mills, the factories were not really a part of the town. They were separate, stood off by themselves, let us say a dozen such mills, each with its village.

The villages, as is usual in America now, not

incorporated. Not to incorporate was a way to disenfranchise men. That had been found out.

On Saturday afternoon perhaps the people of the mill village drifted into the town proper. What a strange staring lot! They lurched through the streets. People of the middle class to whom I had talked had said to me: "I keep off the streets on Saturday afternoons. There are such queer horrid people on the street."

The feeling of superiority to labor had grown and grown with the degeneration of labor. It has become an almost universal feeling among the American middle class.

And now the middle class is being wiped out too. Ask in such towns as the one of which I am speaking here what the coming of the chain store is doing to our middle class.

I was walking in a mill village at night, preparatory to going in, seeing the night shift at work. I had plenty of time, having walked out from my hotel. I was to go in at two. The superintendent of the mill was to meet me at the mill gate at that hour.

I was walking in the mill village and came to

a church. I sat down upon the steps. Thoughts there too. In these mill villages, in many mill villages built about factories everywhere in America, the churches also had come under the control of the employer. So much was this true that in another town, visited during the winter, there had been a strike. A number of employees of a mill, living in such a mill village, were shot. A funeral was to be held and not a minister of the town would officiate. They apparently did not dare ally themselves, even in the death hour, with the strikers, the workers.

It was a miserable ugly little church on the steps of which I sat. Mr. Henry Mencken has already written about the ugliness of churches in industrial towns. I was thinking about the degeneration of men under modern industrialism, of letters coming to me from working men, of factory employees feeling every year more and more out of place in American life.

The churches had gone in for advertising too, for the whoopla. Drums were being beaten at the doors of churches. One minister, an American, had but a few days before addressed an assembly of his fellow ministers, in an eastern city. I had read the account of the meeting in a newspaper. Many thousands must have read it. I myself showed it to several people. They were not shocked, although a few of them did laugh.

The minister, addressing his fellows, had told them of how he had succeeded in getting people into his church. He had gone for a time and had got employment in the advertising department of a tooth paste concern. He had found out how they sold tooth paste. "We have got to use the same kind of methods in selling religion," he had declared.

NIGHT IN A MILL TOWN

AT NIGHT the lights of the mill, standing in the midst of a mill village, are bluish green. The mill dominates the village. It rears itself up. Those who live in the mill village never lose consciousness of it. It dominates the village as the steel mills dominate Pittsburgh, Birmingham and Youngstown, as the great rubber plants dominate Akron and the automobile factories dominate Detroit. There is always the feeling of the mill in the air.

For one thing, there is, in the mill village, always a steady singing roar. It never stops. The mills, when there is no industrial depression, are run day and night. The singing roar creeps along the streets, it invades the houses.

There is a man, a workman, sleeping in yonder house. On the night when I went into the Southern cotton mill and sat, for perhaps two hours, on the steps of a little dark church, I could see faintly several small houses.

There was the workman, in the house asleep, a man of the day shift. He slept heavily but the mill came into his dream.

There would be a little curtainless room, the window partially open. The man sleeping in there would be one of a new class, come into American life, within the last fifty years, with the coming of the factories. He had thrown himself onto the bed, where a man, now at work on the night shift, had been sleeping all day.

The air in the room would be stale. The roar of the mill crept in through the window. It invaded the man's dreams.

Wheels flying in his dreams, belts flying, there would be, even in his dreams, the steady pounding roar of the machines.

The man snores. The snores fall into tune with the roar of the machines in the nearby mill.

A little cry within me. I was once in the mills. I was caught there.

I escaped.

"Have I come back to the mills? What am I doing here?"

That cry within me that night. Memories of my young manhood — in Chicago — mill-caught, money-caught.

That night in the mill village I found myself saying over and over to myself the words of another American writer, Mr. Henry James.

Mr. James had spoken of the passion for success in America. He had spoken of the "bitch success."

A quite elegant man, that Mr. Henry James (a much more elegant and mannered man than I would ever be) and yet he had spoken of success as a "bitch."

"And a good word too," I thought, thinking of money and success as a bitch, my fellow Americans having got scent of her long since, they trotting at her heels, that peculiar look in the eyes, the look of humiliation not real, pride not real, hope not real.

The bitch was trotting in the South now. She had long since invaded my own Middle West, turning men from the cornfields and the forests to the factories and the factory towns.

Making a new kind of man. I gave up that picture. My readers will admit it was not so nice.

I began again thinking of myself as a young man, a factory hand, wandering at night in Chicago streets.

Mill-caught.

Mill-caught.

One of the nameless ones.

Mill-caught.

Mill-caught.

I remember saying to myself that night: "I do not have to go into this mill. If I go in I can come out."

I was reassuring myself—old memories awake in me.

The houses in the mill village were all alike. I had seen the same village in the daytime. A woman had driven me in there and up and down through the streets.

The streets were not paved and as there had been rain we went skidding and sliding through. I have already explained that the village I was in was not a first class one. I had seen others in which the houses were well built, with flowers in the front yard, the roads in front of the houses smooth and hard.

In each house there was a bath room (and toilets had been put in). The houses were well painted. At one mill a landscape expert had been employed. He went about through the village, going from house to house, looking after the flowers, the bushes and trees.

He stopped to talk to housewives. "I think there should be a tree here. We will put it in.

"Here there should be a flowering bush, over there a bed of flowers."

"I think zinnias would be nice."

In one of the mill villages I had heard a tale The owner of the mill there liked petunias. He had a bed of petunias put before each cottage. A friend asked him, "Why this universality, this madness for petunias?" the friend asked.

"But I like petunias," the mill owner said.

It was so in some villages I had visited, but in the one in the midst of which I sat on the night when I went into the mill, there had been no such improvement.

The village was quite new and the factory to which I was going was entirely modern. I had already been inside of the factory in the daytime and now I was to see it at night.

The factory was new and perhaps the owners had not yet had time to employ a landscape man. The village had been hurriedly put up. There had been a hot red South Carolina plain with a railroad running through and there was a river.

Along the river bank there had been trees growing and it was a country of small renting farmers.

Some of the trees had been cut away and the small renting farmers had been gathered in. Men with their families had come down out of the hills. There had been small poor hill farms abandoned, the doors no doubt left hanging open. I had myself seen many such mountain cabins on country roads all through the South.

I had talked to a mill owner about all this. "You see how poor and miserable they were on the farms, these men and their families," the mill owner said. He spoke of life in the little cabins in the North Carolina, South Carolina and Georgia hills.

In many of them there was no food during the winter months, except perhaps hog and corn meal.

A pig would be slaughtered in the fall and a little corn made. The children had no schools, or very poor ones.

They were underfed or badly fed. The men and women of the hills and plains knew nothing of birth control. Children came with alarming rapidity. "Look at the women and girls of the farms and hills and then look at our mill women and girls," the mill owner said. "We have brought them into our villages, it is quite true. Why, you see, we have put up schools for them. Look at our schools. Go into them. Look at the teachers."

In most of the mill villages there is an arrangement made with the county. The mills build the schools —many of them fine roomy brick buildings—and pay half the salaries of the teachers.

They introduce athletics, pay athletic trainers.

They have women experts employed who supervise the food given the children. They build the churches and pay the preachers.

"The mill owners have given to thousands of Southern poor whites a peep into a new life. But," I could not help thinking, "the man is speaking of white labor in the South exactly as men down here formerly spoke of slaves. When the mill man spoke of his white labor something came into his voice that made me think inevitably of an older Southerner speaking of his blacks.

"We have to work them, yes," the mill owner said. "How else are we to pay for all this?

"They know nothing of the modern world, these people.

"They do not know how to feed their children, they know nothing of education.

"You writers speak of us as crowding them in here. It is true we build houses for them. You condemn us because the houses we build for them tend to be all alike.

" 'There is too much sameness to all this,' you say."

As I sat that night in the dark mill village, thinking of all this, I could in fancy see before me the mill owner speaking of his mill village.

He was a man, I thought, as deeply in earnest as most of the reformers to whom I had talked.

There he stood before me, a short strong figure of a man, of the middle age, with a grey mustache and blue rather cold but intelligent eyes. He was a man of the new South, the industrial South. He was one of the leaders of the new age in America.

There was resentment in him too. A little sputter of bitter words came from his lips.

"Who sent you in here to pry into all this?" he asked.

"Me?"

"Yes, you."

"I was asked to come by a woman."

"Ah," he said. "A woman, eh!

"We are being run by women.

"They are the worst of all. It is the women we will have to fight."

At any rate, the man and I were thinking along the same plane there. For a long time I had been convinced that American life had passed into control of the women. I did not resent the fact. It had come about because we men of America had fallen down on our job.

We had come into a new age in American life, had been swept into a new age by the machine and the men in power in American life had no program made for the new age.

We were without statesmen, without leaders.

Our men in positions of power were all secondary men.

"Let the women have it," I had been saying to myself for a long time. As for myself it seemed to me that, in affairs, I was as willing to follow women as men leaders.

The mill owner's mind returned to the accusation that the houses of his mill village were all alike, that there was too much sameness to the life there.

"Why, man, it is the modern note," he said. "It is so in every department of life now.

"You are not a man of the mills," he said, looking sharply at me. "You are a writer. I have heard of you.

"So you go about — let us say you go about.

"Be honest with me," he cried. "You have not complained to me of the fact that the houses in my village are all alike but others who have come here from the outside have complained. I dare say you will do it too if you write of us.

"'Long streets of houses, all alike,' you will write. You will speak of sameness, of monotony."

"Well."

"You came here in a car. It is a cheap car, but it is a good one. I saw you get out of it. It gets you about. Now tell me, how many cars, just like the one in which you ride, are there on American roads?

"Are not the stores in all the towns just alike?

"You read a newspaper. Have not all the American newspapers become just alike?"

The manufacturer who spoke to me of all this was perturbed. Any man who thinks that the successful men of our modern American industrial life are happy men is a fool.

The wealth of America has brought happiness to no one. There is little happiness here. Our early fathers spoke eloquently of the rights of men.

Life, liberty and the pursuit of happiness.

But we did not pursue happiness. We pursued the twin bitches, money and success.

There is no happiness on that road.

There is little laughter in our streets. I myself had gone, in my wanderings, to where the sons, daughters and wives of our industrial giants went to amuse themselves.

They were dreary enough places, American centers of amusements. If a man wanted fire, color, life, it was best not go to them.

"Better go to Coney Island. It comes nearer," I thought.

The man speaking to me of the sameness of the mill villages, himself one of the owners and the builder of such a village, was perturbed.

He was a man singled out among many, all doing the same thing, accused for what all were doing.

There was something weary and disturbed in his voice too.

"Such people as I bring here, to work in my mill, cannot build houses," he said. "They don't know how.

"Of course, I have to build houses for them.

"Am I to build each house as though it were to be my own home?

"To do so would take too much thought," he said. "It would take too much time and thought.

"I can't do it. I have something else to do," he

declared. We were sitting in the office of his mill. "I have to keep this going," he said, making a movement with his arm.

"There are a thousand things to be thought of here.

"I have to buy and I have to sell. I must find a market for my goods.

"We Americans are in competition with all the world.

"We have built up something.

"We have given American people goods, such a flow of goods as was never known in the world before.

"We have set a fast pace.

"The world is trying to catch up with us.

"It is a race," he said.

"For goods then we must give all?" I asked.

"What do you mean by 'all'?"

"I mean freedom, individuality." I remember quoting to the American cotton mill owner the saying of a famous painter, a great modern revolutionary spirit among painters.

"Life can only give you that, the opportunity

to develop your own idiosyncrasies," the painter had said.

The manufacturer laughed at me, a bitter laugh, I thought. "All that is bunk. That's gone," he said.

"Then you believe in this, in what we have got instead, you think it inevitable?"

"Yes."

"Then you are one?"

"I am what?"

"You are a communist."

"A what?" he cried — staring at me.

GHOSTS

I HAD left one manufacturer puzzled by my declaration that he was a communist, but was not America being prepared for communism by all the modern forces of industrial America? Why, how gloriously it had all been arranged. Would it come only after a bitter struggle—through fighting and the terror? I was in a mill village at night. I got up and walked. I walked toward the mill gate. I remember stumbling through dark little streets.

There had been rains. The streets of the village, being unpaved, had been cut and gullied by rains. Dogs barked at me. In the darkness I met what at first I took to be a man walking with stooped shoulders. He had long arms swinging loosely from the shoulders.

In that light he was a grotesque. Did he move or did he merely stand and stare? He seemed an old man with heavy forward-thrusting shoulders. It almost seemed that his head grew, not between the shoulders but out of his breast. His arms were like pendulums swinging at his side.

He stood quite close to me, so close in the narrow street between the houses of the mill workers that it seemed to me I could have put out my hand and touched him.

"He is just a man, an old workman perhaps going home," I thought. He might have been some kind of night watchman. My thoughts had been a little distorted. Now my thoughts, playing me a trick, a trick well enough known to all imaginative people, had made of him a grotesque. He seemed to me of a sudden the symbol of what men had become in the mills, come out of the mills to stand looking at me.

The mill itself loomed up there before us. There was, as there always is to the modern American factory, the suggestion of a vast prison. Nearly all modern mills have high steel fences surrounding them. As industry in America has grown more powerful it has withdrawn itself more and more from the lives of the general.

Mills within the last few years have shown

more and more tendency to withdraw from cities and towns and make their own towns. It needs but a flat place, a river for water, a railroad and cheap labor living close about. The failure of agriculture in America is furnishing plenty of cheap labor.

The labor does not need to be skilled. The day of skilled labor among anything like the generality of workmen has passed. Any kind of strong young man or woman labor will do. Most of the modern machines are automatic. They will do their own work.

Can it be that the instinct, on the part of our industrial leaders, to draw away from towns and cities, to set up separate towns, unincorporated, to build high steel fences about the mills, is an unconscious part of a program to make mill workers beings quite separated from the rest of American life?

What was it about what I at first took to be a mill workman, standing there in that mill village street that night that stirred me so?

He was I thought just a dim rather huge figure of a man, with unusually long arms. He may have been a little stoop-shouldered. The reader will have to bear in mind that I was excited. There were preconceived ideas in my head.

As the figure continued to stand thus I stopped in the middle of the road. My shoes were covered with red South Carolina clay. The man also seemed to have stopped permanently. We were now quite close to the mill gate. There were two tall steel posts with a light at the top of each. As we stood thus, what I at first thought to be a workman and myself, two young girls walked past the mill gate under the light. They came into the light and passed out of the light....

The girls may have set me off. One of them laughed. They both had erect straight young figures. I am sure they could not see the strange figure and myself, standing there in the darkness.

In the figure near me I felt bulk and no strength. For some reason the figure of the Ameriman man, the workman as I thought, seen thus that night, seemed suddenly to represent to me all modern American life.

Was it impotence I felt in him—or It?

Had the word impotence come at that moment into my mind because I had been spending months in mills?

Were the mills making men impotent?

Two girls had laughed by a mill gate. Were

they laughing at the strange figure standing there and at myself?

Why had the word impotence come to my mind?

As for the figure, seen thus in the mill village street, it seemed suddenly to move slowly away from me. That was perhaps an illusion. I also moved toward the mill gate. Recent rains had washed deep gullies in the street. I fell into one of them. I fell forward in the soft red mud. Not only my shoes but also my clothes were covered with the soft red South Carolina mud. When I had gathered myself up and had got to the lights by the mill gate the mud clinging to my clothes was like blood. I cut I am sure a strange and absurd figure standing thus in my muddy clothes, waiting for the mill superintendent to come to the gate and take me into his mill.

. . . .

IMPOTENCE

That word ringing in my ears. When I arrived at the gate of the mill that night, having had that odd experience, having met that figure in the streets of the village — the figure that seemed to me so grotesque — a being having bulk and no strength — the head misplaced — the arms too long, swinging too limply at its sides . . .

The whole thing in some way dead.

Was it a thing human or not human? Had I seen it move?

A thought came to me. "Why," I said to myself, "perhaps it was not a man."

The village in which I had been walking was not very old. The streets had not yet been paved. The village was a place flung up almost overnight, as American towns and villages are always being flung up in our industrial centers.

"Perhaps it was a machine I saw, standing idly thus, a road-making machine," I thought.

I even thought of going back into the village to investigate but looking at my watch I saw it was near the time when I was to meet the mill superintendent at the mill gate. I took a handkerchief out of my pocket and going nearer the light over one of the stone pillars on which the mill gate hung tried to clean my clothes.

I remember laughing at myself and at my own thoughts. As I stood by the mill gate, rubbing away at my black cheap American clothes, trying to remove some of the South Carolina red clay that, in that light looked so like blood, thinking that perhaps I had been alarmed, frightened, upset by a dead machine, a road-making machine standing idly in a street, myself become something grotesque....

I remember a feeling creeping over me. I am not a large man in my own consciousness of myself, although, to some of my friends, I seem at times to appear physically large.

In my own consciousness of myself I am, often enough, a peculiarly small and ineffectual man. "That may be the reason," I say often to myself, "that I write so boldly."

"I am trying," I tell myself, "to find in words a boldness not in myself."

I became a Charlie Chaplin that night by the mill gate. I was, to myself at least and for the time there in the half darkness, just the grotesque little figure Chaplin brings upon our screen.

He, Chaplin, when he was more among us, the little figure with the cane, putting the hat back correctly on his head, pulling at the lapels of his worn coat, walking grotesquely, standing blinking thus before a world he does not comprehend, can not comprehend —

Brushing his clothes, as I was doing with a soiled pocket handkerchief — "he would have been," I thought, "just the one to run as I had done from an idle road-making machine, thinking it a man, his quick rather fragile mind and feeling upset — his eyes distorting things as I so often do."

Himself also trying to put up a bold front to the world.

Myself, in my own consciousness, that night become a kind of Chaplin.

I would myself be half buried in shadows and in darkness. The building before me would be, as I say, dark and huge but lights would shine out from its thousand eyes.

It would be a fact, that building, as the machine is a fact in American life, as the factory is a fact, as the radio and the automobile are facts.

I stand there, small and shrinking, a kind of Chaplin, striving to rub South Carolina red clay off his clothes.

Think of me as a rather impotent figure. Modern male impotence in the face of the machine, as we, in America handle the machine, accepting it and not accepting it, dreading it while we admire, not daring yet to begin to use it as it might be used, for all of us, is the statement I am trying to make in this book.

Be sympathetic if you can but do not spare me, the American artist. We also have failed. We also have become impotent in the face of the machine.

That is what I am trying to say.

And as a kind of touch of what I am driving at, put in the two little mill girls I saw pass the mill gate that evening. Make them almost flappers if you will.

Make them slight and fragile but bold and challenging too. Make them stare at me with something of disdain in their eyes.

Am I asking too much?

I am asking for a statement of the inner strength, of the living potence of present-day American women, of their hunger, the potentiality of new strength in them, that may save American civilization in the very face of the machine.

That, as you will see, if you follow my book, if I succeed in writing it, contrasted with the present-day impotence of the American man.

ENTERING THE MILL AT NIGHT

THE MAN who was to take me into the mill came at last and escorted me in through the mill gate. It was two o'clock in the morning. The gate clanked behind us. He looked at me with curiosity in his eyes.

He had been told I would come there at that hour. Perhaps I had been explained to him. I would have given something to have heard the explanation.

It would have been an explanation offered by the mill owner, that alert red-haired man.

"There are these fellows come here. We have to put up with them," I could imagine him saying.

"They come here and look about. They scribble words."

Alas, it has been going on for ages, this word scribbling—fellows like myself looking about, poking their noses into places.

We have so much less power than they think we have. The drift of life goes on. Our words scribbled are like rain drops in a vast sea. No sensible writing man thinks he has much power.
He is afraid of power. "Power is a disease. I hope I shall not catch it," he keeps telling himself.

"If we do not let them into the mills that will not stop them from writing of us.

"We have nothing to conceal from them."

There would have been something a little defiant in the red-haired man's voice.

"Labor has become what in America it has become.

"How am I responsible? Am I my brother's keeper?

"We men of affairs, we were given the present capitalist system. We did not invent it.

"Let us listen to these Bolsheviks and where will we land?

"It is easy enough for them to talk, or to scribble words as this fellow will do.

"Damn them. Well, you know how I feel.

"However, be courteous to him. Let him look his fill."

The young mill superintendent was not however defiant. I looked at him as he was unlocking the mill gate and afterward as we walked, under a blue light toward the door of the mill. In that light the red mud on my black clothes had become purple.

The mill superintendent was young. He was a strongly built young man with clear untroubled blue eyes.

He was puzzled by my presence there at that hour, by my strange desire to go into the mill at night, to stand in the presence of the flying machinery at night, but he was not defiant.

"He is a young scientist," I thought, my mind leaping away from the fact of the mill, for the moment not hearing the roar and clatter of the machinery in the mill, and for the time centering upon him.

"He is one of the lucky men," I thought, and again the figure of our young American Lindbergh came into my mind.

"They would be, these men," I thought, "fellows with minds centered upon the machine."

There was a whole school of such young men come into American life. You saw them everywhere. They were young men working in chemistry, young inventors, they flew and built airplanes, they managed the mechanical side of factories.

They were clean young men without vices. They studied hard, did things well.

They were men, I thought, walking beside one of them, going from a mill gate to the door of a mill, under a bluish light, thoughts flowing fast in me, they were men, separated by a terrible gulf from others, from older Americans like myself.

They were the figures of the new age, cold men, clear men, impersonal men.

Life, the stormy muddle of life, as I knew it...

The life filled often with distorted dreams, faces of crushed people crowding into dreams...

The sense of defeat in great bodies of people, the slack mouths, the dead eyes...

The world of labor gone down and down—men who worked with their hands having their work degraded...

The work itself made into a mere meaningless mechanical process...

That reacting upon the souls of men...

Making them spiritually impotent...

Physical impotence, perhaps in a whole race of men, to follow that. It was inevitable, I thought...

All my own nature revolted, shocked by the thought.

There are times when these storms break thus upon a man. The result of the thinking and feeling of a whole life may come to a head thus in a few minutes...

Less than that. There are thoughts that cut down into a man like a knife thrust in the belly on a dark road.

The thought I had at that moment had been growing on me for a long time.

It is important to get it simple, to get it clear if possible.

Let us suppose you to be a man who drives an automobile. I do. You drive along a paved road.

You are, let us say, a man in comfortable circumstances in life. You own a good car.

But there is another man, also driving, just ahead of you in the road. He is a broad-shouldered, tanned man, a farmer.

Let us create that man, that scene. There is a long sloping hill.

The man is driving an old Ford.

There are several children in the car, his children.

He is a man who owns and operates a little farm that is not very rich.

He has, however, married, as a young man, and children have been born to him. There are now eight, ten, twelve children. I myself know one poor hill farmer in Virginia who, by one old wife, has had nineteen strong children.

And while we are at it we might as well put the wife also in the Ford. There she sits, surrounded by her brood, a bit shapeless now, her face red and wind-bitten.

These are, to be sure, all imaginary figures. They are, as any man of sense knows, for that reason none the less real.

The farmer is driving the old Ford, slowly, painfully up a long sloping hill.

To be sure, the car struggles and misses fire. It can hardly make the hill.

You, in your fine roadster, have but to step on the gas.

You shoot past. What is in your mind?

Is there not a queer vicarious sense of power in you, of false power?

Admit there is.

There is this kind of vicarious power scattered everywhere through the American world now. Does it seem vicarious power to us because we, as a people, do not own, direct or control it?

It doesn't matter much what you are. You may be a scribbler of tales like myself, a merchant, a lawyer, a doctor, a stock broker.

You have money enough to buy power. It is all arranged for you.

You can even buy it on the installment plan.

Why is it that my son, a young man, yearns to own a car more powerful, more flashy, with more speed than that driven by my neighbor's son?

He wants power, doesn't he, without paying

for it? He wants this vicarious feeling of power...

To pass, like a flash of light, that man, that farmer, with his eight, ten, twelve children in the old dilapidated car?

Why, my son should be afraid of the power he has not earned. It is not power of mind, of feeling. He has not yet trained his imagination so that it reaches out into life beyond the imagination of my neighbor's son.

He has got power to lord it thus over that farmer, a man who has worked all his life in fields, out under the sun and wind, who has raised food to feed many people, who has been all his life honest and simple-hearted, who is the father of all these children, future citizens of the State.

It seems to me now that, as I write, I am drawing near the American danger.

Impotence.

We being the people who have most whole-heartedly surrendered to the machine, without in any way attempting, as a people, to control it.

A malign thing in this light.

How has my son got this machine in which he goes sweeping, with such lordly air, past that man, on that hill?

He has got it, let us say, because I am an indulgent father. I had some money ahead. I wrote a check.

I put that poison into him.

I am sitting in my own car. I step on the gas. It requires but a slight movement of my foot. The car shoots forward.

It goes at terrific speed. It climbs mountains.

The makers of cars understand the feeling that gives me. You will see it implied constantly in the advertisements.

I am to be made to feel superior to my fellows.

Do I accept that feeling as real? Do I take the power in the car to be my own power?

There is something dangerous in all this. It is too easy to surrender to this false, to this vicarious feeling of power.

I am doing it.

You are doing it.

All Americans are doing it.

When I do it I do something dreadful to myself.

Nature will pay me back for this. I cannot go on so, cheating nature.

I am on the road of impotence.

Every claim I make, every feeling I have, of power that is not my own, that is merely bought, is a cheating of the inner me.

It is a kind of terrible self-abuse.

It is the road of impotence. It is a kind of impotence that has been creeping over men for a long, long time. The machine age has but accentuated it.

But what of the young mill superintendent, escorting me into the mill that night? Let us say that men of his type create the power that is so disastrous in its effect on the rest of us. What of them? There has been in me for a long time a great curiosity about such men.

I had seen them everywhere as I had gone about in offices and factories. The young Lindbergh had fixed the type in American life.

They were the men who would be the leaders in life if the Communistic State came.

They would be cold and clear. They would be the young Lenins. "Perhaps the Communistic State has already come more truly to America than to Russia," I thought. These fellows were real evidence of its coming. They fitted in. They believed in power for its own sake. The young communists to whom I had talked were all peculiarly like our industrial leaders. They believed in an end to be achieved and were not particular about the method to be employed in achieving their end.

In the young communist and the industrial leader there was always the same feeling. They were egotists. There was an insane kind of egotism in them.

Young men with cold impersonal eyes doing in government what the young engineers did in the factories. Young men, having been trained for the work, went into the factories seeing and feeling the new age, the age of the machine, as a natural thing. Lindbergh had even tried to personify the machine. "We," he said, speaking of the airplane and himself out over the Atlantic.

"We were in a fog. Hail fell on our wings."

These men, as I had known them, had a sense of something I did not have. They were a new kind of young autocrat. Working in one of the modern great factories such a young man saw the whole factory as one great machine.

It was a machine of many parts. There were bolts, rods, flying wheels, flying belts. There were delicate intricate little parts to all the great modern machines.

Machine making had come to that. It had come to the time of refinement.

Here is a tiny steel spring that releases this valve. The valve lets a certain amount of gas into this cylinder. The gas explodes.

The machine is given a forward thrust thus.

Everything must be coordinated, it must fall into a rhythm. I knew one such young engineer with finely attuned ears. He had an ear developed like the ear of a fine musician and in another age might have been a musician.

He went into one of the great rooms of a factory and listened. There was a singing roar in there, an intense sustained crashing roar. It, like the machines from which the sounds came, was made up of a thousand, ten thousand little parts of sound.

His ear separated the sounds and his mind analyzed them. "There is a machine there not doing its work. See to it."

So finely was his ear trained that he never made a mistake.

The great machines in the room had each its thousands of little parts. When a part was worn, when it did not do its work properly it was scrapped.

It was so with men too. Men had become merely parts of this machine. There is this man here. He must pull this lever. He must synchronize his movements with the movements of the machine.

When he can no longer do it, when his nerves have given away....

He is like the little valve that has become worn, he is no longer of use here.

Do not be carried away by sentiment. Scrap him.

Modern industry is like war. It is war. In war the indivdual doesn't count.

You have to face things, be cold, be clear-headed, be impersonal.

The machine is the new god. "Worship your god, man."

Thoughts flying in my own head thus.

The young man, the young scientists, chemists,

engineers were those who had accepted the new god.

The machine was their God.

It was not surprising that this was so. For some months I had been going into factories. I went wherever the chance offered. I went and stayed in factory towns.

It was a woman who had got me to do it.

She had come to me, to where I was staying, in a small town.

I had escaped from the roar of the big industrial towns, from the cities.

I was in a quiet place.

I was like a turtle with its head drawn in, sleeping under a bush.

The woman had poked me with a stick. She had forced me to crawl out from under my bush.

I can remember the scorn with which she spoke to me.

There were such men as myself. We called ourselves artists. I was of the type.

It was true I was a little old. I had been born

and had grown, at least to physical manhood, in a certain period of American life. It was the life of the whole western world but in America it was intensified.

It was a period of change, of transition. I had written of the period of change. I had been a writer for twenty years.

I had written stories, poems, novels, essays.

Unlike many American writers, when the change in life became apparent, when it became obvious, when mechanical invention followed mechanical invention, the automobile coming, the flying machine, the radio, when the world about became filled with new noises, when speed took the place of purpose, when it became more and more difficult for men like myself to live, I at least had not tried to get out of it all by fleeing to Europe.

I had at least not gone to Paris, to sit eternally in cafés, talking of art.

I had stuck and yet, as that woman pointed out to me, all my efforts had been efforts to escape.

Time and again I had told the story of the American man crushed and puzzled by the age of the machine. I had told the story until I was tired of telling it. I had retreated from the city to the town, from the town to the farm.

So there I was on my farm when that woman came. I had sharply the sense of her laughing at me. To her I was but another man frightened by the machine.

I remember that we walked under some trees while she spoke of it.

She had herself been going to the factories. She was interested in women in factories.

They, the women, were pouring into the factories everywhere, she said.

There were men in the factories too, many thousands of men. The factories were doing something to the men, they were doing something to the women.

"There is a kind of potency in the machine," she said. "It goes on. It does what it starts out to do.

"At least it keeps running until it has quite worn out."

I remember that the woman called me a coward. She said cowardice was the note in American men now.

We went about like children, with our eyes closed, crying.

There were no men here of sensibility, with some human tenderness left in them, going to the factories.

She spoke of our public life. None of the real changes bound to come in life, forced upon us by the machine, was being really discussed by our public men.

It was true there were no scholars, economists and others, who saw into the future. They had however no way of reaching the public ear. They needed what help they could get from men life myself.

If we were not too cowardly at least to attempt to give it.

Writers, she said, men like myself, who should at least be trying to tell the story of my own time, were still in the Victorian age.

We were always writing of love, not understanding love.

We made sex the vital, the important thing in life. The woman said I had done that.

"You think that if you can solve the sex problem you solve all."

She laughed when she said that.

"As though there were no impregnation other than sex impregnation.

"As though life wasn't always all one thing. You couldn't ignore parts of it.

"You had to take everything in."

"You mean we have to take in the factory too?" I asked.

"Surely," she said, smiling.

Or we were hard-boiled men, gone the smart way. We had thrown overboard human life. We thought it didn't matter. We had become that amazing modern American thing, the wise-crackers.

We said that the mass of men did not matter.

There were to be a few supermen, young mechanical geniuses, young engineers.

These men would grasp the new world, the mechanical world, and would control it.

A few great capitalists, controlling the money of the world, a few great chemists, warriors, organizers, engineers . . .

The world was made that way. To be interested in the generality of men was to be absurd, a sentimentalist.

The hard-boiled young artists were brothers to the young engineer, the young financiers.

They were not questioners. They accepted.

"Achieving thereby," the woman said to me, "a shallowness almost inconceivable."

Accepting everything, questioning nothing.

"But what am I to do?"

"Go and look."

"Stay looking."

"Come out of your shell."

"Go to the factories, it is a new age. The new age is to be worked out inside the walls of the factories.

"You who call yourself artists know nothing of the factories.

"They have grown and grown in size, controlling life more and more.

"They control the press now. They control the magazines.

"Plenty of people go into the Ford plant. They do not go with seeing eyes.

"What of the 'belt,' that new god in the factories?

"Who has explained that, what it does to men?

"Why," she said, "there is already a new god. Men have found the new god.

"It is power.

"They worship blindly.

"What about it?"

It was a woman talking to me thus who had got me where I was at that moment. She had got me inside a great modern cotton mill, a cloth mill at night.

I had been there in the daytime.

I went in. I walked in the great rooms. As it was in the cotton mill so it was in the automobile plant, in any one of a thousand plants making every kind of goods.

There were, it was true, still a few badly or indifferently run plants. They are going fast. The big efficient clean orderly places, such as the one into which I had now come, were wiping them out.

The modern industrial world was no place for the muddlers.

PERHAPS WOMEN

THE MAN, the young mill superintendent, and I, went into the mill. There was a little hallway and we stopped for a moment in there. I had the feeling we were staring at each other.

There would be that question in his head:

What does he want here?

Men and women are coming into factories. They are escorted. Such factories as the huge Ford plant at Detroit make a specialty of escorting people through.

They come in, farmers from their farms, town people, merchants and lawyers. Society women come. They walk through in their soft fluffy dresses.

They are in a world of which they know little and sense less and still they are impressed.

The workmen and the workwomen at the machines stare up at them.

Why, there is a world, a life here, of which

those who come thus into the great rooms know nothing. The machines are doing something.

The machines are weaving stockings, they weave cloth, they shape iron. Shoes are shaped in machines.

The visitor sees before him a great machine. Inside the mill all is in order and outside, often, all is disorder. In a certain cotton mill town in the South, at the end of a peculiarly disorderly street, I saw piles of old tin cans along a roadway as I drove down to the mill. There were weed-grown fields and women and men were shuffling aimlessly through the street.

The morning was a dull rainy one. A wife of one of the owners of the mill had taken me there...

Inside the mill I saw a Barber-Coleman Spooler Warper.

It was a machine just introduced into that factory, an extension of the thought, of the imagination, of some man, a machine that threw many men out of work.

The factory superintendent at that place told me it cost twenty thousand dollars.

That was more money than I had ever had. The statement did not impress me.

He said that its introduction into the mill did away with the labor of a certain number of hands.

That statement did not at the time impress me much. The machine is pushing men aside. That is going on everywhere. "Let it," I said to myself that morning.

I stood before the machine. It was a mass of moving parts. Its movements were as delicately balanced as the movements of a fine watch.

It was huge. It would have filled to the last inch this room in which I now sit writing of it.

But can I write of it? I cannot say how many parts the machine had, perhaps a thousand, perhaps ten thousand.

It had Herculean legs.

It unwound thread from one sized ball and wound it onto another. The white balls of thread moved about, up and down along hallways of steel. They were moving at unbelievable speed. As the thread wound and unwound, the balls moving thus gayly along steel hallways, dancing there, being playful there, being touched here and there by little steel hands directing their course, so delicately touched...

So delicately directed....

Bobbins being loaded with thread . . . I dare say bobbins being loaded with many colored threads . . .

Perhaps some silk, some rayon, some cotton.

I may, for the time, have stepped outside the province of this particular machine.

I remember a woman, a mill owner's wife or daughter, tall and delicately gowned, standing near me. I remember two mill girls, one with a mass of yellow hair. No, it was just off yellow, with streaks of gold in it . . .

Her fingers were doing things rapidly, with precision. I did not understand what she did.

Dancing balls.

Dancing rods.

I remember thinking rebel thoughts, to me new thoughts.

I must have stared at the woman who brought me there and at an alert blue-eyed mill superintendent.

Thinking of artists, striving blunderingly, as I am doing here, to express something.

No accuracy to their movements — if they be

writers no words coming from under their flying fingers with such beautiful precision.

There, in that machine, what seemed at first disorder in movement becoming a vast, a beautiful order.

Why, a man goes a little daft.

A thousand, perhaps in the life of such a machine a hundred million, white balls, each containing to the hundredth part of an inch, the same yardage of slender thread...

They dancing down steel hallways, every hop, every skip calculated, they landing at little steel doors, never missing...

They being touched, handled, directed by fingers of steel.

Never harshly to break thread that I could break easily between my two fingers.

Thread flying, at blinding speed off one spool and onto another.

These handled, something done to these. In this shape, this form, they are serving some obscure purpose... in this vast modern passion of goods making.

I am describing this particular machine in a room far away from it, in a quiet room, no technical description of the machine before me, the accuracy of my description mattering nothing...

An impression sought, something beautiful, something in movement beautiful.

Something in tone beautiful, in sound beautiful.

Why, there is power here. Here is the almost god.

A crazy new grace—

Steel fingers jerking—in movements, calculated, never varied...

Great arms moving...

Materials touched with such delicacy of touch as I can never know.

I remember standing in that place, that time. I shall never forget that.

I remember thinking of men of my time, thoughtful men, earnest men, who would have destroyed all machines.

I remember there had been such thoughts in me.

I think it must have been the vast order in the mass of steel parts, all in movements, that had caught and held me so.

I, all my life, a lover of artists and their work ... men working at least toward order.

Thinking — "these men who designed and built this machine may some day be known to be as important in the life swing of mankind, as the man who built the Cathedral of Chartres.

Whispering to myself —"They may be the real artists of our time.

"We in America may be, unknowingly, in one of the great forward-thrusting times of the world."

Thinking also of that woman standing there beside me as I looked at the machine, it in some new way exciting me ...

A sardonic thought. I am sure I said no rude words to the delicately bodied, delicately gowned rich woman who brought me into that mill.

I thought suddenly, staring hard at her.

"Hell," I remember thinking, "you are a woman delicate and lovely, but you will never find you a lover who will touch that body of yours with the delicacy and strength with which those white balls of cotton are being touched."

Thinking:

"Is blood necessary, is flesh necessary?"

"We humans are but little bundles of nerves. Our nerves betray us.

"We think we think.

"In the machine we have made a thing infinitely more masterful than ourselves."

It was a moment of pure machine worship. I was on my knees before the new god, the American god.

Looking up again at that woman standing there.

"You have to wait for hardness in your lover, if you have a lover.

"Here is always hardness.

"Here is always the thing done, accurately and truly.

"No blundering here."

Myself not hysterical, not made hysterical by the wonder of that particular machine...

I have felt dimly the same vast order sometimes in the stars, walking at night on some country road.

I have felt it in rivers.

I have felt impotence too. This is not a feeling individual in me. I challenge any painter, songmaker, word-arranger, any poet, to go stand where I stood.

A Barker-Coleman Spooler Warper in a cotton mill will do. It is enough.

Why, if he, the artist, had made that machine...

Let him stand as I did, not having made it, never in his whole life having made anything that moved forward, doing its work, with perfect order...

Never having loved perfectly, created perfectly...

Let him be a workman at such a machine...

The man, the workman, does little but start and stop it.

It works outside him...

I, a man, can go blunderingly into blundering other lives.

I can fail in the eyes of others, as I will fail in this book, trying as I am here to say the unsayable.

I can fail because you who read fail also.

Your whole life is a story of failure.

As for myself, all of my success as a writer has been in telling the story of failure.

I have told that story and told it well because I know failure.

The machine does not fail.

I ask you men who read to follow me.

Ask yourself...

"What will it do to me, as a man, to stand, pulling a lever, let us say, to a machine that does not fail?

"Can man, being man, actually stand, naked in his inefficiency before the efficient machine?"

Men, you know it cannot quite be done, not yet in any event.

We know this—impotence comes from the fear of impotence.

In our machine age how can we help fearing?

Why, I was in an American Cotton Mill at night. There was a mill superintendent with me. I think I ought to tell you, who have not been in such a mill, either in the daytime or at night, a little of how cotton from the farms is made into thread and then of how, in the great loom rooms, it is woven into cloth.

The cotton mill is a complex thing like all modern mills. It has been built up slowly from small rude beginnings. Here is this cotton, brought into the mill in its bales. It comes from the fields.

There is a story there too, the story of Southern cotton fields, but it cannot be told here.

In the mills the machines begin to handle the cotton. They roll and toss it. Now it has begun to move forward in the mill, a moving snowy mass.

As it moves forward the machines caress it, they stir it—iron fingers reach softly and tenderly down to it.

The cotton has come into the mill still impregnated with the dust of the fields. There are innumerable little black and brown specks in it. Tiny particles of trash from the fields, bits of the dry brown cotton boll, cling to it, tiny ends of sticks are enmeshed in it.

The cotton gin has removed the seed but there are these particles left.

The fibre of the cotton is delicate and short.

Here is a great machine, weighing tons. See the great wheels, the iron arms moving, feel the vibrations in the air now, all the little iron fingers moving. See how delicately the fingers caress the moving mass. They shake it, they comb it, they caress it.

Every movement here is designed to cleanse the cotton, making it always whiter and cleaner, and to lay the delicate fibres of the mass, more and more, into parallel lines.

Why, this cotton is already on the road to becoming. It is becoming goods. It moves with roaring speed toward that end.

Long months spent making this cotton in the fields. All the danger of bad weather, boll weevil, drought.

Hope coming, despair . . . a farmer's whole family spending months making a bale of cotton. See how nonchalantly the machines eat it up.

And now it is clean and has begun to emerge from the larger machines in a thin film. You have been in the fields in the early morning and have seen how the dew on the spider webs, spun from weed-top to weed-top, shines and glistens in the morning sun. See how delicate and fragile it is.

But not more delicate or film-like, not more diaphanous than the thin sheet now emerging from yonder huge machine. You may pass your hand under the moving sheet. Look through it and you may see the lines in the palm of your hand.

Yonder great ponderous machine did that. Man made that machine. He made it to do that thing. There is something blind or dead in those of us who do not see and feel the wonder of it.

What delicacy of adjustment, what strength with delicacy! Do you wonder that the little mill girls—half children, some of them—that the women who work in the mills—many of them I have seen with such amazingly delicate and sensitive faces—do you wonder that they are half in love with the machines that they tend, as modern boys are half in love with the automobiles they drive?

But we are in the weaving room now. It is another huge room. The room is a forest of belts. The belts, hundreds of them in this one room, go up to the ceiling as straight as pine trees in a Georgia wood.

They are flying, flying, flying.

In the loom room, visited that night, there might have been fifteen hundred, perhaps even twenty-five hundred looms, all in the one great room. This mill had many thousand spindles. The looms are not so large. They come up to a man's waist.

They clatter and shout. They talk like a million blackbirds in a field. Here, in this room, as everywhere in modern industry, there is something vibrant in the air. The inside of such a room is like the inside of a piano, being played furiously. It is like the inside of an automobile, going at eighty miles an hour.

If I could make you feel this. There is wonder and terror in this room. The night accentuates it.

The whole story of labor in modern industry is a story of nerves. That I have found out. It may be the story of all modern life.

Can man get on top of this? Can man retain the beauty, the wonder, the efficiency of these modern mills and not be destroyed by them?

It is obvious they have destroyed old ideas of government, of the relationship of man to man.

Had I anything in common with the mill superintendent who walked with me that night?

There were the men who made the cotton in the fields and the mill hands in the mill.

They had little or no sense of each other. That is one of the tragedies of modern industrial life.

The workman in the furniture factory has no sense of the lumberman cutting trees in the hills.

The man in the steel mill does not sense the miner. In the South men come from the cotton fields to work in the mills but they lose there, almost at once, the sense of fields.

It is the machine that does it. The machine has become a wall between man and man. One of the striking things about the modern labor world is the loss of a sense of a common interest.

The mill superintendent went smilingly along. His nerves seemed unshaken. I had oddly the feeling, that night in the mill, that he thought me a little silly, and perhaps affected, to be so moved.

WILL AMERICA HAVE TO TURN TO WOMEN?

THIS little book is a record of thoughts, of feeling in the presence of something amazing in modern life — the machine. It is difficult to get continuity. I have a desire to write down in words the thoughts and feeling of a man walking beside a mill superintendent through a modern factory, going from room to room, now falling in behind my escort as we walked in narrow aisles between rows of spindles, now trying to speak to him, to appear calm, to appear casual.

The spindles were flying at such speed that all sense of motion in them was lost. I remember speaking to the man. "They are standing still," I said. I had to shout my words. He replied in a low voice. He had learned the trick of so pitching his voice that it carried clearly to me through all the roar.

His voice found a little crevice in the great roar of sounds. It crept clearly through to me.

He laughed.

You can carry speed to that point — to the point where there is no sense of movement.

You can carry the sustained roar of the mills to that point, to where there are little layers of silence.

You can carry modern life in the presence of the machine to that point — to where there is no sense of human life —

To where thoughts stop, motion stops —

To where there is no life, outside machine life.

To impotence.

The mill superintendent's voice went on, explaining things to me. My nerves tingling.

The reason they run the mills at night it seems is this — well, all this machinery costs tremendously. Suppose you have an investment of a million dollars in machinery.

A million dollars, it seems, cannot stand idle. It must work, work, work.

Men like myself, who will never understand

finance, cannot comprehend this. If someone gave me a thousand dollars, ten thousand dollars, a million dollars, I would lay the dollars aside in a heap, I would think of them as so many dollars lying there, waiting to be spent.

But it seems money isn't like that.

Money is power. Power must be used. The mill costs too much. It cannot stand idle.

Idleness would destroy it. The cost of the money that bought the machinery would consume the machinery.

There is something very complex here too, a thing called finance.

The machinery must run, run, run. It must work, work, work. People must run the machines.

Night does not matter. Time exists during the night too.

In the presence of some of the most powerful of modern machines it seems that time does not exist at all. More than once I have stood before such a machine for an hour, not knowing I had been there beyond three minutes.

As a man stands sometimes before some wonder in nature or as a man stands looking into the eyes of his beloved.

But there is something very strange about the night. To be, even as a guest, in a steel mill, a shoe factory, a stocking factory, a cotton mill at night is something different from being there in the daytime.

Why, even the mill village had become something new and strange at night.

Night is the time of love, of strange thoughts, of dreams.

See Pittsburgh or South Chicago in the daytime and then see these places again at night.

At night even familiar streets of towns and cities become sometimes strange. In the daytime when you are in the mills there are often windows open. You, who are walking in the mill, are carried, let us say, into the very heart of the new age, into the new industrial age.

You are in the new age of movement, of speed. You are carried forward in this river of power but you are not submerged in it.

When you are in one of the great mills in the daytime you walk to a door or to a window and look out.

There is perhaps a little town to be seen. A

woman goes along a street. She is a fat woman and has a market basket on her arm.

Or there is a dog trotting across the mill yard.

Or you see distant fields, distant patches of woodland.

The old life of man outside there, the sense of an old life comes back — before the belts began to fly.

— before the time of flying wheels

— before the time of the river of goods pouring out to men — submerging them perhaps

— too many clothes

— too many automobiles

— too many pairs of shoes

Speed, speed —

Power subject to man.

Man subjected to power.

Goods.

Goods.

Too many goods.

PERHAPS WOMEN

THIS little book is intended to be a statement. It may be an absurd statement. It is that modern man cannot escape the machine, that he has already lost the power to escape.

He has lost the power to escape, because what the machine can do to men has already been done. Man has already accepted the power given him by the machine, this vicarious power that moves mountains, that flies beneath the sea and through the air, that transports him so swiftly from place to place, as real power. He has accepted it as his own power. He is like a man who has indulged for a long time in self-abuse. He can no longer stand erect.

You do not need to take my word for this. The impotence of the modern man is felt everywhere. Why, it was man, the male in the world, who formerly produced our leadership but now there is no real male leadership. To attain real power, of the mind, of the spirit, is a long slow process. Why should man go to all this trouble when he can so easily attain this vicarious power? He can attain it by getting money. He doesn't need much. The ownership of any sort of an automobile — and who so poor he cannot in America own a car — the ownership of any sort of an automobile will give him all he is asking, this fake feeling of power.

Formerly men created the poetry of the world, the religions of the world, but there is little or no poetry being produced now. Religion, in the old sense, is practically dead.

We are in a stalemate. Everyone feels it. Shall we have to turn the American world over to women? I think we shall.

I think it is time now for women to come into power in the western world, to take over the power, the control of life. Perhaps they have already taken it. There is plenty of evidence that they have.

To be sure there are all sorts of women, but we should not be confused by that fact. That there are plenty of silly women in the world means little.

The world has always been run by leaders in any event, and it seems to me that the new leaders must be women . . .

Because, as I have already tried to point out, the woman, at her best, is and will remain a being untouched by the machine. It may, if she becomes a machine operator, tire her physically but it cannot paralyze or make impotent her spirit. She remains, as she will remain, a being with a hidden inner life. The machine can never bring children into the world.

"Nor can women," someone says, "without the assistance of the male."

Well, there is the rub. There is where our hope lies. If these machines, brought by man, so casually into the world — they, the machines being what they are — such amazing, such beautiful manifestations of man's imaginative power — they at the same time having this power to destroy man, if these machines are ever to be controlled, so that their power to hurt men, by making them impotent, is checked, women will have to do it.

They will have to do it perhaps to get men back, so that they may continue to be fertilized, to produce men.

THE CRY IN THE NIGHT

THERE is a cry going up out of present-day men to the women but there is a cry coming from the women too.

I heard it that night when I went to visit the factory.

The machines in that factory were doing their work. I was caught up fascinated by what was going on in the room into which I had come, as I have always been caught up, swept out of myself, by what I have seen and felt in modern factories.

I was in the factory that night and thoughts went on in me. Perhaps I had gone a bit daft. The machinery in the great room was going at terrific speed. The night also helped to make it a strange world into which I had come.

I had gone to stand by myself on a little raised platform, the young mill superintendent having left me there, when an odd thing happened. There was an accident. The lights in the great room in which I stood suddenly went out. The room was plunged into darkness.

There was no stopping of the machinery. It was, as I have explained, a room in which cloth was woven. The machine in there went on weaving cloth. How long they could have continued to do it, I don't know.

I know that I was lifted into a new, into a queer world. I had been watching the work people, the operators in there. As in every mill into which I had gone, in every large office where both men and women are employed, I had sensed something still alive in the women that seems to me going dead in the men.

The women in the factories, in the offices, and the shops are still alive. They are not enervated spiritually by the machines. They have not accepted the vicarious feeling of power, got from machines, as their own power.

The power of women is more personal. It is a matter of human relations. It operates directly on others.

It is a power the machine cannot touch.

In the factory that night, when the lights went out, I stood trembling on the little raised platform to which I had climbed and tried to stare down into the roaring darkness below. There were people, workers, men and women, down in there. The darkness in the mill lasted but a few minutes.

I have already told of how voices can carry through the terrific roar of the mills. The voices find the little crevice in the sustained roar. There were voices that night.

It began with a woman's voice. She laughed hysterically, I thought. It was a young girl's laughter. "Kiss me," she cried. Was she calling to the machines? Machines do not kiss. She laughed again. "Kiss me while the lights are out," the voice said. A male voice from far across the room answered, wearily I thought.

The male voice was not much. "Who? Me?" it asked, wearily.

"No, not you."

There was a little chorus of male voices.

"Me?"

"Me?"

"Me?"

"Me?"

(hopefully)

"No, not you. None of you."

"I want a man," the girl's voice said. It was a clear young voice.

There was an outburst of laughter from many women — ironic laughter it was, down there in the darkness — and then the lights came on again.

The mill was as it was before. It roared on. Men and women workers in the room were staring at one another. "The women often do that sort of thing," the young mill superintendent afterward said to me.

"Why?"

"Oh, they are making fun of the men," he said coldly enough.